Seahorses

Mysteries of the Oceans

www.bunkerhillpublishing.com

First published in 2004 by Bunker Hill Publishing Inc.
26 Adams Street, Charlestown, MA 02129 USA

10 9 8 7 6 5 4 3 2 1

Library of Congress Cataloging in Publication Data available from the publisher's office

ISBN 1 59373 039 X

Designed by Louise Millar

Printed in China

Seahorses

Mysteries of the Oceans

CATHERINE WALLIS

BUNKER HILL PUBLISHING

BOSTON

for Holly, Anna and Jack

Introduction

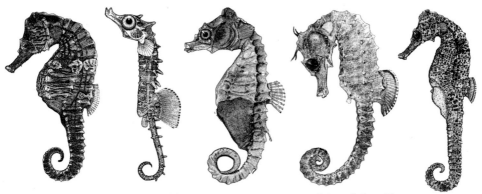

Seahorses in all shapes and sizes. From left, Hippocampus erectus: erectus's *juvenile form;* H. zosterae; H. guttulatus; H. reidi; *and* H. erectus *with weedy camouflage.*

The seahorse is one of nature's most startling creations. For the ancient Greeks and Romans, who found them washed up on shore after storms, the only explanation for such an astonishing form was a mythological one. The ancients believed that in the deep, seahorses grew to the size of real horses and pulled the chariots of Neptune. They could be glimpsed from afar, leaping on the crests of the waves.

The seahorse's appearance, even to us nowadays, seems out-of-this-world. The creature is a fish, but has the head of a horse and wears a crown. It has a tail, but a curly one like a monkey's, and instead of scales a suit of armor plating. Its eyes swivel monstrously in two different directions at once, and it sports a pot belly like a beery old man. In a most unfishlike and contrary manner, the seahorse swims upright, its progress slow and ponderous. Yet its fins are as delicate as a fairy's wings and move so fast they are almost invisible.

What is the sum of these extraordinary parts? For one so tiny, sometimes hardly bigger than a fingernail, the seahorse is full of surprises. Despite the earnest expression and shy tilt of the head, it is a voracious eating machine, deadly to small shrimps. Each one of its curious features bends toward a single purpose: survival. Its solution to the thorniest fish problem of all – how to protect eggs in open water and ensure the survival of the next generation – is unique: male pregnancy.

A seahorse may live out its whole life within three square feet of seaweed, seagrass, or coral reef. To balance the fact that it can't swim very fast and escape predators, it has perfected the most artful camouflage. But there are dangers it cannot protect itself from. Seahorses thrive by the millions in their underwater forests all over the world, but they are also caught and sold by the millions, for a purpose little known in the west: the creatures are dried, boiled, and marinated for use in traditional Chinese medicine. China has a population of over a billion people and the belief in this medicine is ingrained. It is a fact that cannot be legislated against, or wished away.

There is, however, a stealthier but more present danger to these beautiful and vulnerable creatures treasured as symbols of the sea: the pollution and destruction of their delicate habitats.

The tiny seahorse in the vast ocean

A seahorse is a feeble swimmer, helpless against the swell and tug of mighty ocean currents. Its best strategy is to anchor itself to something – a branch of coral, perhaps, or a piece of weed – and call it home.

Seahorses are not adventurous. Sometimes the male, whose primary job is to raise the babies, will stake out a single piece of seagrass or seaweed and make it his permanent address. He may venture only

This nineteenth-century engraving shows an emerging coral island, a refuge for seahorses and all manner of marine life in the trackless ocean.

A seahorse fossil from the Lower Pliocene period, eight million years ago, found in Rimini, Italy. Photo by kind permission of Raimund Albersdoerfer.

three feet or so away in any direction and spend his whole life in this small circumference, while the female travels at least ten times as far in search of food. But compared to many fishes, seahorses are stay-at-homes.

They tend to live in coastal areas on the fringes of continents or groups of islands, where there is more light and therefore vegetation to hide in and hold onto – places like coral reefs and seagrass meadows, estuaries and mangrove swamps. They even adapt to man-made jetties and piers. Some live on quite open and featureless stretches of the seafloor, disguising themselves against predators by adopting a suitably drab color.

Globally, these quiet creatures are found in almost all the seas and oceans of the world, except the very coldest in the north and south. They appear to have evolved about 40–50 million years ago. Seahorse fossils show that they have remained virtually unchanged in this time.

How have they been distributed so far and wide? Many species have "pelagic" young, which means that as hatchlings they swim straight up to the surface and spend the first few days – or even weeks – there while they grow. The current moves them along, and eventually they descend to new homes in new places. Sometimes they will attach themselves to

A Victorian book-spine: although so tiny, the seahorse has long symbolized all that is remarkable and enigmatic about the ocean.

A long-snout (H. reidi) *holds tight.* Photo by Tony White.

Seahorses encircle the globe in tropical, subtropical, and subtemperate regions.

clumps of floating seaweed which carry them far off to new territory.

On an evolutionary timescale, their wide dispersal is probably due to dramatic earth-shaping factors such as changing sea levels and continental drift. The same or very similar species can sometimes be found thousands of miles from one another.

There are somewhere between 35 and 50 species worldwide, ranging in size from $1/3$ inch to nearly 12 inches (16mm to 300 mm), the majority being about 6 inches (150mm) long. New discoveries are still being made, and the seahorse map has yet to be fully drawn.

The elusive seahorse

Seahorses are notoriously difficult to find in the ocean, let alone identify. In the nineteenth century, when there was a fever of discovering and naming species, about 120 kinds of seahorse were described from new specimens all over the world, with much repetition. *H. kuda*, variously known as the Estuary, Oceanic, Common, Yellow, Spotted, Crowned or Topknot Seahorse, was identified under eleven different names. The taxonomy of seahorses therefore ended up in a complete muddle. Through the efforts of modern-day seahorse lovers, it is being put in better order. Currently, one authority puts the number of species at 35, another at 55.

A mere glance in an aquarium tank containing a single species will illustrate why these little creatures are so tricky to identify. Within the same group, one can be dark brown and another yellow, or red. The size of their spines varies. Each will have its own array of skin filaments, its own stripes and spots. Males and females vary in slenderness and spininess. Seahorses even differ according to mood – a resting one can be much more subdued in color than an active one.

Hippocampus kuda *was named in 1852 by Dutch fish expert Pieter Bleeker, who compiled his famous* Atlas Ichthyologique *(1877) after many years spent in Java studying marine life. He is credited with identifying over a thousand fish species in his lifetime.*

13

A page from the journal of John White, written while on an early sea voyage to Australia and published in 1790. The seahorse is given a general label, "hippocampus," and is drawn swimming horizontally, like the fish above it. Not for another sixty years would different species of seahorse start coming to light from every corner of the globe.

The most accurate way of telling different kinds of seahorses apart is to laboriously count the rays that support the fins, count the bony rings on the trunk and tail, and compare snout lengths and head spines. Generally speaking, these vital statistics are predictable from species to species.

Information about seahorses is patchy, and extensive data exists for only a few species. Some that have been positively identified have scarcely even been seen, let alone scrutinized. *H. minotaur*, the Bullneck Seahorse, for example, was named in 1997 from specimens trawled at a depth of about 200 feet (60m). We know that the Bullneck is roughly 1^1/$_2$ inches long (40mm) and has no spines, which may mean it is adapted to live on sponges. But no one has yet found it alive. *H. fisheri* was discovered in 1903, but so far only three specimens have been found, one from a dipnet offshore, one from the stomach of a dolphin, another in a trevally. In a century, there have been no live sightings.

Unknown species are undoubtedly living quietly undiscovered. It is reckoned that only about one-tenth of one percent of the ocean has been thoroughly explored. Australia's Great Barrier Reef alone, which is better combed than most areas, has 2100 individual reefs, not to mention 540 islands

Patterns in nature repeat uncannily: this seldom-seen Zebra Seahorse (H. zebra) of Australia has solved an underwater camouflage problem in almost exactly the same way as its land-dwelling namesake.
Photo by Rudie Kuiter.

with reef fringes, for seahorses to hide in. The Indo-Pacific region is largely unexplored but has already turned up the greatest variety of these little creatures.

15

In May 2003, a pygmy seahorse spotted in a photograph by PhD student Sara Lourie, was hunted down in the Flores Sea of Indonesia and miraculously found. It was, as she suspected, a separate species, and she named it *Hippocampus denise* after the underwater photographer Denise Tackett, who took the photo. Perfectly camouflaged for Gorgonian (fan) coral, it is the tiniest seahorse yet discovered – $1/3$ inch long (16mm), about as big as a fingernail.

The Low-crown Seahorse (H. dahli) of Australia uses several different forms of disguise, including zebra stripes which make it hard to distinguish from the Zebra Seahorse (previous page). Only vital statistics like numbers of rings and tubercles finally decide which species is which.
Photo by Rudie Kuiter.

Opposite: Hippocampus denise, *so far the tiniest addition to the growing list of seahorse species.*
Photo by Denise Tackett
© Tackett Productions.

Seahorse Lineage

On our planet there are about 1.7 million species known species so far, yet the oceans are almost entirely virgin territory. The Census of Marine Life, which is conducted internationally, has been attempting to track down all the life forms in all the seas and oceans of the world. There are 210,000 species on the list, with hundreds of thousands more to find. Of fish alone, more than 100 new kinds turn up each year.

Seahorses belong to a large and widespread family of fishes called pipefishes (Syngnathidae), the vast majority of which have long thin bodies resembling the weeds and grasses they live amongst. In all species it is the father who nurtures the young on his own body.

The modern system of classification began in the eighteenth century, devised by the botanist Linnaeus, but the relentless addition of species has meant that groups and sub-groups have been added or rearranged to a confusing degree, so that every reference on seahorse classification is slightly different. As it currently stands, the family tree of our *Hippocampus* (Greek for "horse sea-monster") looks like this:

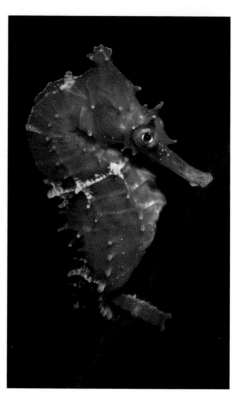

*Eight new species of seahorse from the Western Pacific region have been named in the last two years in Australia. This High-crown Seahorse (*H. procerus*) is from Queensland.* Photo by Rudie Kuiter.

Classification

KINGDOM: Animalia

PHYLUM: Chordata

SUB-PHYLUM: Vertebrata

CLASS: Actinopterygii (fishes with jaws and bony rayed fins)

INFRACLASS: Teleostei (about 25,000 species of bony fishes)

ORDER: Gasterosteiformes (9 families, including stickle-backs, tube-snouts, trumpet fish, pipefish and seahorses – many with elongated snouts and body plates in the skin.)

FAMILY: Syngnathidae (known generally as pipefishes, and having about 330 species)

SUB-FAMILY: Hippocampinae (sea-horses and pygmy pipehorses)

GENUS: *Hippocampus* (our seahorse)

Carolus Linnaeus, eighteenth-century creator of our modern system of classification.

Seahorse Relatives

After a century and a half of discovering species, natural historians have created four sub-families in the Syngnathidae clan. The largest sub-family are the true pipefishes, long, slender fish that range between 4 and 8 inches (10 and 20cm), with a few much larger or smaller species. The tails have no

Seahorses, despite their markedly different shape, are members of the pipefish family.

caudal fin and are often prehensile, like the seahorse's. There is a second, much smaller sub-family of pipefishes with tail fins shaped rather like ping-pong paddles.

A third small family consists of 13 species of pipehorses and seadragons. Seadragons vie with seahorses for attention because they are so extraordinary looking, and so fantastically camouflaged. They are too delicate to keep easily in an aquarium, and are difficult to spot on a dive in their native South Australian waters, and therefore have the kudos of being a rare visual treat.

Seahorses and pygmy pipehorses make up the final sub-family. The pygmy

This light-hearted monograph was published in 1958, one of the first books for ordinary readers devoted solely to seahorses and their curious relatives.

A small group of pipefishes called Doryhamphinae *or Flagtails have distinctive paddle-shaped tail fins.*

Two seahorse relatives, a bellowsfish and a pipefish, are delightfully given the names Sea Woodcock and Tobaccopipe-fish in this nineteenth-century engraving.

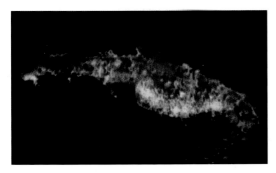

A Short-pouch Pygmy Pipehorse (Acentronura breviperula), *the seahorse's closest relation. Photo by Rudie Kuiter.*

A nineteenth-century depiction of a Weedy Seadragon (Phyllopteryx taeniolatus).

pipehorse may represent an evolutionary link between the pipefish, which swims horizontally like a normal fish, and the seahorse, which swims upright. Rudie Kuiter in his book *Seahorses, Pipefishes and their Relatives*, observes that a pygmy pipehorse's modest pouch causes it to swim at a slight incline. With an even heavier pouch to bear, the seahorse is thrust into a vertical position, its head bent for balance, and in this curious pose we begin to forget it is a fish at all.

Leafy Seadragons, Phycodorus eques, *found only in Southern Australian waters. These creatures resemble seahorses but swim horizontally and are from the separate sub-family of pipehorses and seadragons.* Photo by Tony White.

Seahorses and seadragons normally turn away from the camera and are caught in profile, but this seadragon obliged the photographer with a full view of its beautiful face. Photo by Tony White.

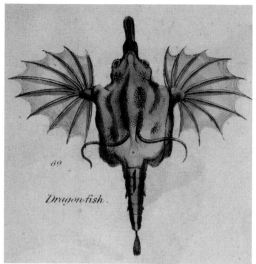

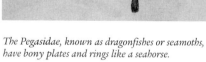

The Pegasidae, known as dragonfishes or seamoths, have bony plates and rings like a seahorse.

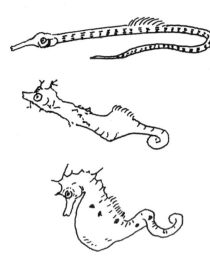

The pygmy pipehorse may represent a stage of pouch development and posture that lies somewhere between a pipefish and a seahorse.

There are several other groups usually included among seahorse relations, though they may only have tenuous similarities. Closest, and from the same order, are the ghostpipefishes, of the family Solenostomidae.

From other orders or sub-orders come bellowsfishes, flutemouths, shrimpfishes, trumpetfishes and seamoths, all of which share a passing resemblance and some morphological features.

Camouflage

The word "camouflage" was coined in 1917 during the Great War, when airplanes were first used for scouting and troops on the ground found themselves with nowhere to hide. Looking to the animal kingdom for ideas, the soldiers adopted protective coloring and patterns to break up outlines, and adorned themselves with bits of foliage. Like animals, they had to find a way to disappear under the very gaze of the enemy.

All wild creatures are perpetually at war, and each has found some advantage to give them the edge over predator or prey. Neverending is the list of cunning devices nature throws up to improve the chances of survival: stings and poisons, decoys, electric shocks, sonar, and sharp teeth, to name a few. The seahorse has none of these. It possesses instead one of the most complex and cunning ploys in the animal kingdom – the ability to change color.

Think of color change and we picture the chameleon, a land reptile which has become the poetic metaphor for someone who changes identity to match the moment. In under two minutes, a chameleon can transform a dull uniform green into the dappled dark and light shades of leaves in sunshine.

Although far removed from one another, the forces of evolution have solved several problems for the seahorse and the chameleon in a remarkably similar way – a flattened head, separately rotating eyes, a prehensile tail, and the ability to change skin color and pattern.

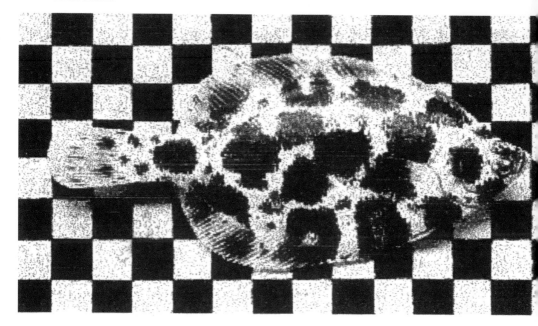

A flatfish attempts invisibility on a checkerboard, and all but succeeds.

Under the sea there are even more exciting examples. A cuttlefish, for instance, can change color in under a second, and change again, and again, to create a continuous and brilliant rainbow wave along its body, all for the purpose of mesmerizing prey or attracting a mate. Common bottom-dwelling flatfishes like plaice and turbot are able to imitate any sand, gravel, or pebble seafloor they happen to be lying in. It's easy to overlook the complexity of shading and speckling required to blend in perfectly: the hand of an artist could not do better. If you place one of these fish over an illuminated checkerboard, it will acquire dark and light squares distinct enough to play a game on!

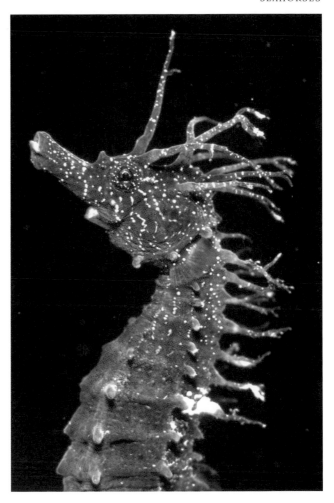

Below:Seahorses, like this H. kuda, *can grow skin filaments in a matter of days to camouflage themselves in a new environment.* Photo by Tony White.

27

There are at least 250 varieties of pipefish of all spots and stripes. Photo by Tony White.

Seahorses have apparently infinite capacity to mimic their surroundings, by looking like a drab bit of vegetation on a muddy seabottom, a piece of bright fan coral or sponge, or a strand of seaweed. The results can be very beautiful. The Painted Seahorse, *H. syndonis*, from Japan, can turn itself from mud-brown to richest crimson or brilliant saffron yellow. The Long-snout Seahorse, *H. reidi*, from the West Indies, can summon an enormous range of colors to match any of the countless bright sponges or corals it lives amongst. Seahorses of almost all species can produce polkadots, lines and stripes, speckles, and Pollock-like splashes to blur their outline.

This is magical but not magic. There are special cells distributed throughout the skin called chromatophores, which contain pigment granules. The granules can disperse, which makes the cell darker, or they can aggregate in the center of the cell, which makes it lighter. In addition, the chromatophores can move a little farther from or closer to the skin's surface. There are muscle fibers attached to chromatophores which can squeeze and flatten the cell to rearrange the pigment granules.

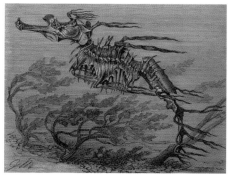

"All skin and grief," this eerie nineteenth-century engraving was probably drawn from a dried specimen of a Weedy Seadragon covered in filaments.

The Mossy Seahorse can produce long straggling tendrils or a fine mossy coating like this little H. japonicus.
Photo by Rudie Kuiter.

This capacity of color cells to move around, expand, and contract means that colors can combine – for instance, black and yellow make green, and orange and blue make brown – and complex patterns can be created. Ultimately, instructions come through visual stimulation of the eyes or direct action of light on the skin, or through hormones. The mechanism is entirely reflexive.

A seahorse can not only change color but can lengthen its spines and grow skin filaments that resemble algae or seaweed. The Mossy Seahorse, *H. japonicus*, lives in muddy, weedy terrain, and can grow a coat to match – long trailing tresses like seaweed, or fine short filaments evenly covering its entire body, like moss. A seahorse that finds itself in a mat of sargassum seaweed will grow the necessary skin appendages to camouflage itself within a few days. Likewise, if it is put in an aquarium where the camouflage is no longer needed, it will lose them just as fast.

The Pygmy Seahorse, *H. bargibanti*, lives in the West Pacific, from Japan to Australia,

on soft Gorgonian (fan) coral, and has found a symbiosis with the coral itself, which becomes incorporated into its skin. The skin reacts by developing knobbly growths, and together with the right colors and patterns, this tiny creature, just $^3/_4$ inch long (2cm), becomes virtually impossible to see.

Taking top prize for cunning is one of the seahorse's close relatives, the Leafy Seadragon, *Phycodurus eques*. Protruding from its sea-horse-shaped body and tail is a fringe of skin flaps elaborately branched and lobed like seaweed. The effect is so convincing that if it stays still, the seadragon is a piece of seaweed – until it swims away.

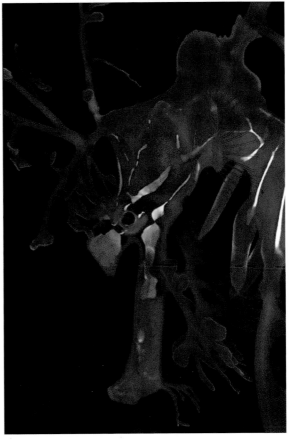

Opposite:The Pygmy Seahorse (H. bargibanti), only $^3/_4$ inch long (2 cm), confounds the eye by exactly mimicking its fan coral home. Photo by Tony White.

One of the most stylish examples of camouflage in nature, the branched lobes of an Australian Leafy Seadragon. Photo by Tony White.

31

Eating . . .

However gentle a seahorse seems, it is driven by a gargantuan appetite for food. It eats mainly tiny crustaceans, and needs in the neighborhood of 50 to 300 of them an hour, depending on size and other factors. It also eats fish larvae of all kinds, and crab larvae and worms, anything small enough to fit down its slender snout.

The reason for this nonstop eating is that the seahorse has no teeth. Food is swallowed whole. Additionally, it has no actual stomach, and the alimentary canal has very poor digestive powers. So there is a constant pressing need for food to keep up energy.

Many aquarium fish shops refuse to sell seahorses except by special order because they know that the impulse buyer, seduced by the seahorse's "cuteness," will soon be fed up with having to feed them three or four times a day. Until recently, when frozen mysid shrimp became readily available, a seahorse owner not only had to breed live bait to feed the seahorse, but culture algae to feed the bait, truly a labor of love.

In the wild, the seahorse uses stealth and ambush to catch their prey. Clinging onto a piece of seagrass or some other anchorage with its tail, it keeps two wary eyes out for

The seahorse (H. kuda) uses its snout like a straw to suck in its helpless prey at lightning speed.
Photo by Tony White.

*A Moluccan Seahorse (*H. moluccensis*) from Indonesia lies in wait, ready to ambush.* Photo by Rudie Kuiter.

crustaceans, snapping at them as they pass within range. Each eye can swivel in independent directions, spotting one shrimp aft and another fore. The seahorse can shimmy up and down a piece of weed like a man climbing a drainpipe, picking off the crustaceans that settle on it. Seahorses will stalk prey, too, threading their way through the fronds with a style both rapacious and elegant.

The only evidence that the prey has been caught is a slight upward jerk of the head, and sometimes, a faintly audible "snap." The cheeks are inflated, and the snout acts like a siphon, sucking the prey straight into the stomach. Seahorses have no teeth and can't chew, so to get a shrimp up the snout, they sometimes have to smash it in two with a guillotine-like action of the jaws. It is not uncommon for them to fill themselves up to capacity, yet like the gluttons they are, they keep on eating, repeatedly spitting food out.

33

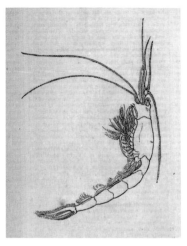

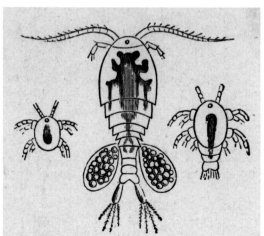

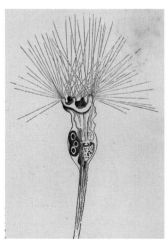

Favorite seahorse plankton greatly magnified: from left, Mysid shrimp, brine shrimp, copepods, and rotifers.

It's hard to see what a seahorse is actually eating because its food is so small. The ocean is a soup awash with invisible or barely visible victuals. Trillions of single-celled algae (phytoplankton) are consumed by tiny zooplankton, who are eaten in turn by small fishlike seahorses. The most numerous animal on earth is a zooplankton that seahorses love: the copepod, a beautiful iridescent crustacean measuring a mere $1/10$ inch long (2mm), or less. It constitutes 70% of all zooplankton in the oceans, yet we are almost wholly unaware of its existence. One copepod alone will consume tens of thousands of phytoplankton an hour, and a seahorse in turn will devour as many as 3,000 copepods or more in an active 10-12 hour day.

A diver on a reef is often conscious of a continuous crackling all around him. This is the sound of thousands of tiny crustaceans making their characteristic snapping noises. The seahorse, far from elegantly wafting its life away among the weeds, is caught up in a noisy underwater world crowded with creatures of every size, eating, and being eaten.

An innocuous stand of weed covered with tiny invertebrates makes a banquet for seahorses.

. . . and being eaten

The sea is a dangerous place, even for the unpalatable seahorse. There is very little meat on this creature – it isn't much more than skin stretched tautly across bone. Hard and spiny, it is hardly the first choice for many predators.

But despite their amazing ability to match their surroundings and pretend to be a piece of seaweed or coral, seahorses do get eaten. Fishermen have gutted them from the stomachs of skates, cod, sea-perch, snappers and tuna. They are eaten by remoras, flat-

*A Thorny Seahorse (*H. histrix)*, not much more than skin and bone.* Photo by Tony White

Frogfishes, which sometimes eat seahorses, can beat them at their own game of camouflage and ambush. Photo by Tony White.

heads, anglerfishes, frogfishes, some turtles, and the very small Fairy penguin in the southern seas. They have even been found inside sharks.

At night, groups of seahorses will assemble partway up a stand of weed to keep away from the bottom and out of the clutches of enemy crabs. An anemone will sometimes pull in an unwary seahorse with its tentacles, and a few days later spit out the skeleton entirely cleaned of its scant flesh. Some fish, like the electric ray, have an electromagnetic sensory system that can detect impulses given off by seahorses' muscles when they move, in which case no amount of camouflage can save them.

Baby Short-head Seahorses (H. breviceps) *cling to a bit of floating seaweed in the first few days of life, a snack-in-waiting for passing predators.* Photo by Rudie Kuiter.

A baby Zebra-snouted Seahorse (H. barbouri) must bid farewell and face life without parental help.

The most dangerous time in the life of a seahorse, as with any fish, is when it is young. Babies are born perfectly formed miniature horses, but they are only around $^1/_3$ to $^2/_3$ inch long (10mm or 15mm), less in the pygmy varieties. They are mere will-o'-the-wisps, with tails the thickness of a thread.

Many species have pelagic young, which means they start off living in surface waters instead of at the bottom of the sea. As soon as they emerge from the pouch, they are instinctively attracted upwards by the light. There they gulp air into their swim bladders, the device most fish use for regulating their level in the water. They may drift in open water for several days, or longer.

The dangers are unimaginable. There is nowhere to hide and nothing to hang onto, except each other. (Indeed, they get into appalling, sometimes fatal tangles, with five or six clinging to each other's tails and noses in an unravelable knot). In this horribly exposed position they are spotted by seabirds or scooped up by fast-swimming surface feeders like tuna. If they are not pelagic but grow up near the bottom, crustaceans and anemones may eat them. It takes time for their pigmentation to develop, so they cannot camouflage themselves. Yet by some miracle, five or six from a brood of 200 or 300 will make it past the baby stage, and nine months or a year later will have achieved a much safer adult life.

The seahorse's curious body

Holding a live seahorse in the hand for a second, one has the impression it is lighter than it should be. It is covered with skin instead of scales, with the usual mucus film that fish need to protect themselves from infection. The skin is tightly drawn over the bony plates and spines.

The horse's head and coronet are so graceful and Etruscan, it is hard to keep one's mind on scientific business. The "neck" is actually a narrowed section of the abdomen. The head can only nod up and down and cannot pivot from side to side. The tilt of the head seems to contribute to overall balance and steering: to ascend in the water, a seahorse will usually stretch its head upward and uncoil its tail. To descend, it will curl its tail forward and bend its head downward.

However strong the horsy resemblance, the seahorse's long snout is certainly not a nose. It consists of elongated facial bones fashioned into a kind of straw, which is used to suck up prey. At the end of this snout is a tiny mouth as vicious as a mousetrap. It opens and closes the snout, and is also used to chomp overly stout prey in two. The seahorse and all its Syngnathid relatives (which

The seahorse derives its shape from an intricate bony architecture. Engravings from a monograph on seahorses by M. Rauther made in 1925.

means "fused jaw" in Greek) have similar tubular mouths. There are no teeth.

Adding an air of importance to the seahorse's noble head is a coronet. Quite what it is for, we don't know. On every individual **39**

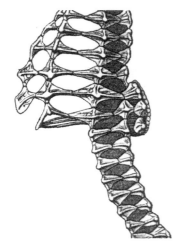

Bony plates interlock to form a cage around the body.

There is also a theory that the coronet increases the size of the male, and the bigger the male, the more desirable he is to the female because he can take on board more eggs.

The seahorse has no neck; its head, bent at an angle to its body, provides balance. Photo by Tony White.

it is a little bit different, so it may be a means of seahorses recognizing one another. The males don't fight with it, and it isn't used for catching prey, but there is speculation that it acts as a sensory device. It also seems to be used to make a clicking or popping noise; a rasp inside it is rubbed against another bony segment, rather like a cricket rubbing its legs together. These clicks may be some form of communication so that seahorses can find each other in the dense weeds and dim light. The sounds increase at the time of mating and, in some species, while eating.

40

The eyes

As a rule, fishes' eyes are much larger than those of land creatures, so that they can gather as much light as possible in the murky depths. They are set on the side of the head and look in different directions, giving a good range of vision. In seahorses, the eyes are exceptionally large, and they protrude, like a chameleon's, and have muscular control that enables each one to turn in different directions. This gives them almost 360-degree vision. Seahorses see in color and appear to be very sensitive to small changes of light and form.

Smell and taste and the feel of water disturbances are all detected by means of sensory cells on the skin. There is no outer ear, but an inner ear which can detect the sounds that carry so much better underwater than in air. When we swim we have no inkling of the cacophony of underwater life because our middle ears contain air which blocks the sound.

The fins

It would be easy to miss the fact that seahorses have fins at all, so gossamer thin are they and so rapidly moving. Seahorses have no caudal (tail) fin which normally provides swimming power. Instead, the dorsal fin propels the seahorse slowly forwards. Two

The delicate rays of the dorsal fin oscillate from side to side so fast they produce a continuous wave, propelling the seahorse forward.

small pectorals sit very high up behind the gills, like a pair of ears: they are used to hover and deftly steer through weed and coral. There is also a tiny anal fin which is probably vestigial.

High-speed camera studies show that the individual bony rays that make up the fins oscillate from side to side as fast as 70 times a second, like a series of metronomes, each slightly off-beat with the next. This creates a wave of the fin that pushes against the sea and achieves motion, but it is a mere flicker to the human eye.

The delicate dorsal fin sometimes gets torn, but will usually grow back in two or three weeks.

The tail

Proffer a finger and a seahorse whose tail is free will naturally curl around it. The grip is surprisingly strong, like the gentle squeeze of a small child's hand, and in the larger species it can even bruise. The tail has an inbuilt desire to seek an anchorage, and when seahorses swim together they constantly grasp each other's tails and snouts, and then irritably throw each other off.

The ocean has layers of water ranging from fast moving and turbulent near the surface, to utterly still at great depths. Seahorses live in the shallower waters of the continental shelf, and though conditions are often calm enough to allow them to swim freely, they need the tail when the current is strong or they would exhaust themselves struggling against it. Storms are thought to be the main cause of death.

In the courtship waltzes that seahorses do together, the male and female "hold hands" (tails) and pirouette. If the end of a tail is damaged or broken off, it can regenerate like a lizard's.

The armor plating

Most bony fishes have thin overlapping scales to protect their bodies against abrasion and smaller predators. But some fishes have developed thick bony plates within the skin instead. The drawback of this form of protection is that it makes it difficult for the creature to bend and move.

In a seahorse the plates are sharply outlined beneath the skin, giving it a distinctive geometric form. The plates interlock to completely encircle and protect the organs, and they bristle with knobs and spikes like a medieval suit of armor. There are roughly a dozen rings round the torso and anywhere from 33 to 48 forming the tail. The variations in the exact number of these torso and tail rings are sometimes the only way of telling species apart.

A pretty engraving of 1869 shows the tail curling backward, but its prehensile tendency is to curl forward.

Opposite: The dorsal fin of a Queensland Seahorse (recently named H. queenslandicus) *is barely visible as it swims through the water. Photo by Rudie Kuiter.*

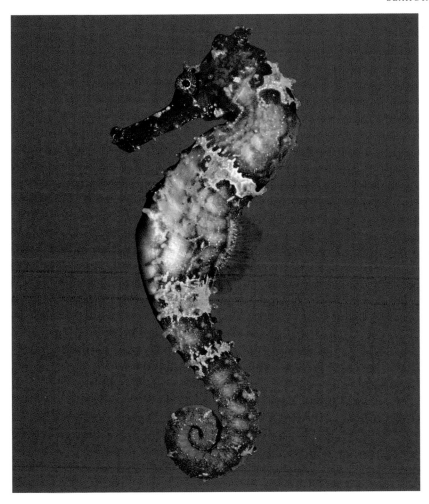

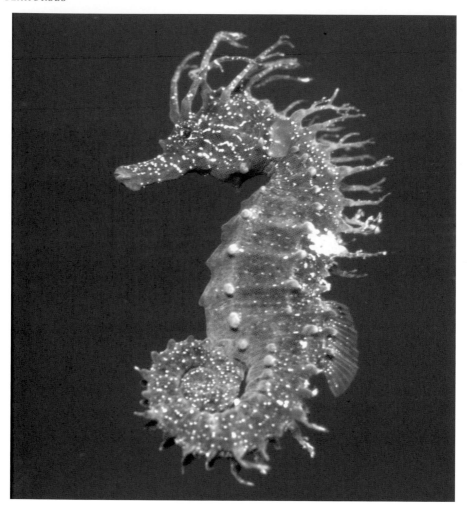

The gills

Like other fishes, the seahorse breathes through gills, but gills with a difference – not the familiar comb-like arrangement, but small tufted gills, like little powder puffs. They open to the surface via a small hole, which is covered with the "operculum," something like a pot lid, that rythmically opens and closes. There are a number of fish with these simple gills, most of them in the same order of Gasterosteiformes (see Seahorse Lineage).

* The pouch is described in the chapter on male pregnancy.

This 140 million year old fossilized ichthyosaurus bears an uncanny resemblance to a seahorse.

Opposite: Its tail tightly coiled in, H. kuda hurries away, keeping a wary eye on the photographer.
Photo by Tony White.

Though dried seahorses as curiosities now seem a frivolous waste of life, in Victorian times their unique bony bodies were the subject of fantasy display cases.

45

The private life of a seahorse

The problems of procreating in a trackless ocean are monstrous. How to find a mate in the murky depths, let alone attract one? How to make physical connection in the restless currents? Where to lay the eggs in all that water? How to protect them from being eaten before they hatch, and after they hatch?

There seem to be almost as many novel solutions to these problems as there are fish. There is a shark that slashes his chosen female with special teeth in a perverse bid to attract her, skates that bite her pectoral fins, the clown fish that trembles before her. There is also no end to egg-laying innovation. The female cardinal fish fashions her eggs into a sticky ball, and the male fertilizes it and rolls it about in his mouth for a week until the eggs hatch. Perches make long jellied egg "ropes," and scorpion fish make a "balloon" out of air bubbles and eggs, both of which are difficult to get a mouth around. The stickleback male makes a clever nest out of weeds and secretions from its kidneys, and then he guards it, fans it, and fusses over it. A Siamese fighting fish will battle to the death any interloper who dares approach his nest.

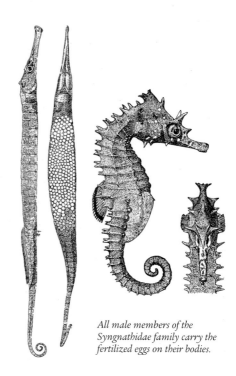

All male members of the Syngnathidae family carry the fertilized eggs on their bodies.

Seahorses, always original, make no exception when it comes to reproduction. The male seahorse has evolved his own built-in nest, a pouch with a vascular lining astonishingly like a womb.

*Australian Pot-bellied males (*H. bleekeri*), keen to fill their pouches, in hot pursuit of a lone female.*
Photo by Rudie Kuiter.

*A male Eastern Upside-down Pipefish (*Heraldia nocturna*) carrying his brood of eggs, about eighty of them.* Photo by Rudie Kuiter.

This complex and remarkable feature solves several problems in one stroke. Fertilization takes place safely inside the body of the male seahorse instead of out in the water. The eggs are completely sheltered. And the most dangerous stage – when eggs hatch into larvae and are most liable to be eaten – is passed in peace and quiet while the babies become fully formed.

In all species of the Syngnathidae family, the father nurtures the young. The eggs are usually stuck to the underside of the body, or covered by a flap of skin, which is a great improvement upon eggs free-floating in water. But evolution has improved the design still further in seahorses and pygmy pipehorses, which have a closed pouch like a marsupial. There the babies grow in blissful safety, entirely hidden from the world.

Courting

Before the pouch can be filled with eggs, a mate must be wooed. Courtship is a matter of timing in fish – male, female, ready-to-fertilize eggs, and the right sea conditions must all come together at once. And in the case of seahorses, the male's pouch must be in a ready state as well.

As always, the seahorse manages to inject beauty into nature's necessities. White's Seahorse (*H. whitei*), from the environs of Sydney in Australia, has been observed performing a nine-hour courtship dance, male and female twining and twisting, sinking and rising together balletically.

They meet, and in each the color brightens at once. They find a single strand of seaweed or some other holdfast, and clinging on with both their tails, begin an elegant

Opposite: After hours of courtship dancing, a White's female (left) transfers her eggs into her male partner's pouch. Photo by Rudie Kuiter.

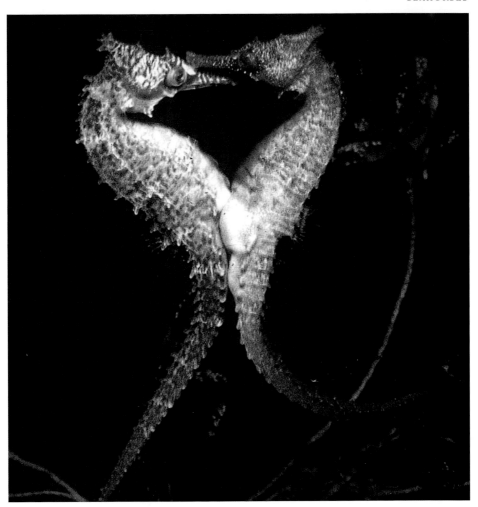

A baby High-crown Seahorse clings delicately to life. Photo by Rudie Kuiter.

roundelay. From time to time, they twine tails and dance along the bottom together, then return to the maypole, then waltz off again. It is a mesmerizing sight that can put a diver into a trance.

As the dance draws to its end, the female begins to tip her head upwards and rise, stretching out her tail and pointing her whole body in a sign of readiness. The male takes the signal and starts jack-knifing his tail up and down in order to pump water into his pouch and force it wide open. He points his head upwards and now they rise in unison. The female has an "ovipositor," sometimes almost as long as her snout, and if the two can coordinate their movements in the current, she puts the ovipositor into his pouch and releases her eggs. The eggs are fertilized on the way in. Mission accomplished.

Monogamy

Seahorses are said to be monogamous, but this is a generalization. It is true in the case of White's and a few other species, but an equal number have many partners. However, that monogamy occurs at all is very unusual, especially among fish. It may help with the entire reproductive process for a pair to be well synchronized. The female has a constantly growing spiral of eggs inside her, and as one batch becomes ready, she pumps those eggs full of water and must lay them within twenty-four hours or they will simply spill out of her, wasted, onto the seafloor. If she readies the eggs when the male's pouch is also ready, none will be lost or wasted.

In the case of White's, the couple greet each other religiously each morning, and perform a five- or ten-minute version of the courtship dance, holding tails and twirling. Then they part company for the rest of the day, which is devoted to eating. This daily intimacy seems to be an essential part of staying together. Experiments have been done in captivity which show that if a male partner is removed after receiving eggs, and replaced by another, the female will perform the morning greeting with the new male, and when the original mate returns, she will shun him for the male she has been meeting daily.

It has also been observed that if a mate dies, its partner may never take another, but this could be because seahorse populations are quite small, and it's simply difficult to find a replacement.

Seahorses have not been found to fight for territory, but in some non-monogamous species, the males will fight for a female. Male Drab Seahorses (*H. fuscus*) from the Red Sea will tail-wrestle and punch each other with their snouts, the loser darkening in color. Little *H. zosterae*, the Dwarf, of the Florida coast, will fight fiercely for his female.

The truth is that we still know very little about the social and breeding habits of these beautiful creatures. They must be studied in the wild, rather than in the constraints of an aquarium, to know their true behavior. There is no shortcut. Garnering information requires many diving hours. Out of a possible fifty species, there is next to no information about at least fifteen, and some are known only from fishing trawls and a chance photograph. Of the rest, only about eight have been intensively studied.

Male pregnancy

The male seahorse's pouch is one of nature's phenomena. Two pockets of skin on either side of the body are joined in the middle, leaving only a small opening. Inside is a lining rich in blood vessels, like a womb. The eggs are fertilized on their way in, and once they are safely inside, the pouch lining begins to change dramatically, becoming swollen and spongelike as the blood vessels enlarge and multiply.

The eggs embed themselves in this soft tissue, which forms a nest around each individual. To make more room, the pouch begins at once to form an inner wall that increases surface area. The large Pot-bellied Seahorse (*H. abdominalis*) of New Zealand grows five of these walls for its big brood of up to 700 eggs. An average figure for a medium-sized seahorse of 4 to 6 inches (10 to 15cm) is about 250 eggs, but in fact the number varies hugely. The Longsnout (*H. reidi*) and the Drab Seahorse are both about

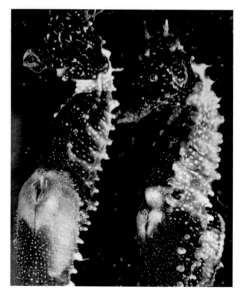

The pouch of a male seahorse seals the young in a womb of perfect safety while they grow.

This baby was born prematurely, and has the burden of a yolk sac to carry with it.

6 inches long (15cm), but one commonly produces 1600 eggs and the other as few as 30. Little *H. bargibanti*, the Dwarf, can sometimes only manage two.

Heavy with eggs and a swollen pouch, the male remains close to its home patch. The inside of the pouch is alive with growth and chemical changes. It produces, for instance, prolactin, which is the hormone that triggers milk production in women. In this case it stimulates the secretion of enzymes. The egg's shell splits open and the embryo grows, still attached to the pouch wall, gradually using up its yolk.

Of course not every egg makes it to this stage; some find no room and simply degenerate. There can be "complications" too. An embryo can die and putrefy in the womb, building up gases which send the poor father floating helplessly to the surface.

But all being well, the little curled-up embryos continue to grow in their separate nests. If the father is monogamous, his mate will visit him every day during the pregnancy. Meanwhile, the fluid inside the pouch gradually changes from a nutrient-filled broth to something like the composition of seawater. This helps acclimatize the babies to the outside world.

When the babies are ready to hatch, the male writhes and contorts himself,

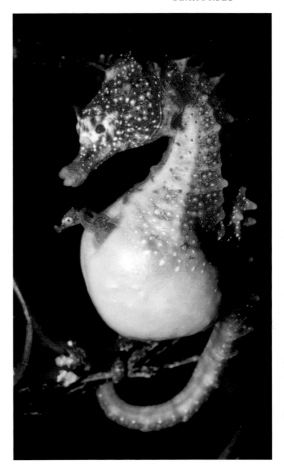

After nearly a month in the pouch, a baby Short-head is born. Photo by Rudie Kuiter.

53

Newborn babies tend to swim horizontally at first.

A Weedy Seadragon, like a pipefish, carries ruby red eggs beneath its tail. Photo by Tony White.

expelling his brood in little batches, along with placental debris and lots of bubbles. He sometimes presses against an object to help push the babies out. His breathing is labored, his eyes wide and moving compulsively with the effort.

The whole birth process can take several hours, or even days. Immediately afterward, the pouch begins to prepare itself for the next brood, and within three or four days the nest is filled again.

Gestation from start to finish takes roughly three weeks to a month, and breeding seasons range from two months to eight, or longer. The Tiger Tail of the Philippines and Malaysia (*H. comes*), which is monogamous in the wild, can breed year round, even in the heavy rains. A faithful pair can efficiently keep on producing babies, month after month.

Of course seahorse broods are small compared to many other fishes, which commonly lay hundreds of thousands and sometimes millions of eggs, then swim away. Of these enormous batches, only a tiny percentage will survive. Fishes that care for their broods tend to have far fewer eggs, but a higher percentage grow to maturity because they have been nurtured and protected for longer.

Among the seahorse's close relatives, the pipefishes, there are some species which protect the young even after they are born. If they have skin flaps on their abdomen rather than a fully enclosed pouch, the father will hover nearby so that the babies can scurry back inside the flaps when they sense danger. But seahorse babies, once out of the pouch, must fend entirely for themselves.

Home

If all the water from all the seas and oceans of the world could vanish for a moment, we would be astonished at the scene revealed – ocean basins far deeper than the land is high, mountain ranges, chasms, volcanoes, giant stalagmites, and as far as the eye could

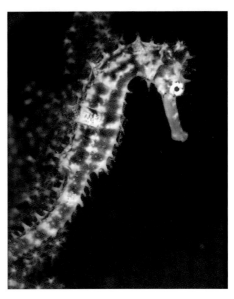

A Thorny Seahorse living on coral that has colonized the WWII wreck the USS Liberty, *off Bali.* Photo by Tony White.

The Common Seahorse (H. taeniopterus), often found on the edges of sunlit seagrass meadows in Indonesia and northern Australia. Photo by Tony White.

see, flat deserts of grey ooze, the abyssal plains. Where could a little seahorse find to live in this frightening terrain?

The answer is that nearly all marine life inhabits the narrow strip of ocean that borders land – the continental shelf. This flood-

Coral reefs grow around volcanic islands and form atolls in the ocean.

ed edge of the continents extends roughly 46 miles (75km) out to sea, and is an average of 500 feet deep (150m), after which the floor slopes sharply down towards the basins. On this shelf, sunlight can penetrate, and algae and reefs can grow and anchor, providing a home base and source of food for millions of other creatures. The small attract the large: all the great commercial fisheries are here.

Beyond lies the deep ocean, plunged in darkness and mystery, subject to phenomenal pressures, and virtually empty. Yet there is life to be found everywhere humans have searched – in the organic mud at the bottom, and even in the Pacific Ocean's Marianas Trench, which lies 7 miles (11 km) below the surface. But it is a lonely place compared to the bustling continental shelf.

Seahorses have been trawled at depths between 200 to 325 feet (60 to 100m) at most, so as far as we know, they are confined to the continental shelf. There is a tendency to find in any one area a species living very shallowly in weeds or seagrass beds or in rubble near the shore, and another species living deeper, on reefs of various kinds. Some move seasonally between shallow and deep.

Seagrasses and seaweeds

A seagrass meadow can be an eerie sight, stretching monotonously into the underwater distance. In reality they are alive with invertebrates and crustaceans – crabs, lobsters, burrowing worms, urchins, starfish, conches, and anemones, as well as young fish doing a bit of growing in a protected place, and seahorses.

Seagrasses grow in shallow sunlit seas, having evolved from land grasses. Quite unlike seaweeds, they have underground root systems, flowers, and pollen. So many crustaceans live on each blade of grass that these meadows are ready-made seahorse pastures. Yet it is a challenge to find them here. Not only are they color-camouflaged but their slow rocking-horse style of swimming matches perfectly the gentle wafting of grass in the ocean current.

Seaweed is equally useful as shelter and pasture to seahorses. Of thousands of species, one of the most fascinating is sargassum, a brown algae equipped with air-filled floats. In the Atlantic, it forms enormous mats that drift over the surface of the ocean. They are most famously found on the Sargasso Sea, which lies over abyssal plains three empty miles (5000m) of water below, forming castaway islands for literally hundreds of species. Some mats grow to the size of football fields, or even larger, and are alive with copepods, crustaceans, and other invertebrates. Many hatchlings take shelter here while they grow, including four types of turtle. Bigger fish like tuna, dolphins, and swordfishes cruise underneath the mats in search of food. There are grave dangers in such a jungle. A seahorse that joins the fray is able in a matter of days to grow skin flaps to disguise itself in the sargassum. These remarkable floating habitats have no doubt helped, in some small part, to disperse their seahorse passengers a little farther round the globe.

Reefs

Because they are so hard to spot, it's exciting to find seahorses on coral reefs, and divers feel a sense of great satisfaction when they do. Reefs are one of the seahorse's chief habitats. Thirty percent of all known marine species live in these environments, clinging to every square inch of coral, living on top of one another, and inside of one another. Seahorses are just one of the weird and wonderful creatures that crowd a reef. Here, uniqueness spells survival.

All inhabitants of a reef are knit together, forming a tightly interdependent community. Coral polyps and their symbiotic algae build the stony edifice. Parrotfishes break off pieces of it, eat the algae and grind the remains to sand, and seagrasses take root in the sand. Molluscs, sponges, sea worms, and algae bore holes in the coral, and all sorts of tiny animals encamp there. A myriad of fishes, colored and camouflaged as if for a carnival, eat the unwary who venture out of their holes. Some, like seahorses, are active from dawn till dusk, while others come alive at night: there is 24-hour action. Animals like starfishes eat the detritus. A fish called the cleaner wrasse offers a special service, picking parasites off bigger fishes, who in return deign not to eat their groomers. The saber-toothed blenny cleverly pretends to be

57

a cleaner wrasse, and then takes a chunk out of its gullible prey.

This world of stony pinnacles, forests, caves, and crevices answers the seahorse's two greatest needs – somewhere to hide and something to eat. At least half of all seahorse species live some of the time on reefs. Some are highly specialized: the Pygmy Seahorse (*H. bargibanti*) seems to have adapted to breed exclusively on a single species of Gorgonian fan coral called *Muricalla*. It incorporates the coral organism into its skin in order to look more like it. Most seahorses on a reef cling to algae, grasses, or corals that don't sting. But Barbour's (*H. barbouri*) of the Philippines and Northern Indonesia

Reef inhabitants live on top of one another – sometimes literally: a nudibranch takes a ride on a pipefish.
Photo by Tony White.

appears to be impervious to the stings of stony corals.

What is remarkable is that the largest, most complex coral reefs grow in warmer waters, which are poor in nutrients. The fewer the resources, the greater is nature's resourcefulness. Reefs are like the suqs of Cairo or the ghettos of New York: every individual, including the seahorse, has to constantly adapt and fight for its little corner of survival.

Sponge reefs

Usually found in the deeper waters of the continental shelf, these reefs are fabulous roving sculptures, sometimes hundreds of miles wide, and they are important to seahorses. Sponges are ancient and rudimentary animals with no tissues and no organs, consisting merely of an open network of cells stiffened by hard "spicules." A sponge will take hold on the bottom of the ocean and grow as a mound. As they multiply, they build up their fantastical bodies in tubular, branching, and lumpen shapes, providing homes for countless creatures within their body cavities. They tend to expand in the direction of the current, and when they die, remain intact, becoming a base for further growth. There are more than 7000 species of sponge, with evocative names like the pink slippery sponge, the spiny tennis ball, and the small tufted pear. At least 11 species of seahorses are known to live among them.

The weird and wonderful bodies of sponges offer homes to seahorses and myriad tiny creatures.

*The Estuary Seahorse (*H. kuda*) hunts among the sunlit shallows of a channel in Indonesia.*
Photo by Tony White.

Estuaries and mangroves

While most marine life lives along the continental shelf, the majority of that is concentrated on its very edges, right up close to the land. Estuaries and mangroves are bursting with life, and many seahorses have adapted to this difficult environment where seawater and freshwater mingle.

Estuaries can be visually rather boring, but for the animal kingdom they provide safe haven from the sea, a place to spawn away from the waves and currents, and to glut upon the newly hatched, before swimming away to the wide ocean. Estuaries are a breeding ground for some crustaceans, which in turn lure seahorses.

There are big problems to overcome in this terrain. In rainy seasons there can be too much fresh water. Estuaries are tidal, so seawater is in constant advance and retreat across the shallows. Animals can get trapped and left behind. Breeding in an estuary requires careful timing.

Similar problems exist in mangrove swamps, where dense mud accumulates in coastal areas and is colonized by specialized trees. These mangrove trees are adapted to exclude salt, and have long thin roots which arch from the trunk up through the air, absorbing oxygen before descending into the dense oxygenless mud. In other parts of the swamp, where the roots are surrounded by seawater, they become encrusted with algae, worms, mussels, oysters, anemones, barnacles, crustaceans, and clinging on with all the rest, seahorses. Many of these creatures can adjust their internal chemistry to deal with the freshwater that runs off the land.

Six seahorse species live at least part of the time in estuaries or mangroves: the Lowcrown of Australia (*H. dahli*), the Shortsnout of the Mediterranean (*H. hippocampus*), the tiny Dwarf of the eastern coast of the United States (*H. zosterae*), the ubiquitous *H. kuda* with its multitude of common names and found from the Maldives to Japan, a possibly new and unconfirmed (and as yet unnamed) species found in the estuaries in Vietnam, and the Giraffe Seahorse from South East Africa (*H. camelopardilis*). Some of these use mangroves and estuaries seasonally, wintering in deeper waters.

The Spinach Seahorse or *H. knysna*, however, is found exclusively in three South African estuaries: the Knysna, Svartlei, and Keurbooms. It has adapted to concentrations of seawater from virtually fresh to 59 parts per thousand (the average for seawater is 34).

Man-made homes

Seahorses are not fussy about what they cling to. They are quite partial to piers and jetties, where there are ropes and old machinery and endless junk in the water to hide amongst. Their camouflage skills make them as inconspicuous here as anywhere, and their only other requirement is food. They also inhabit "rubble reefs" which have been built as protection to harbors and inlets against the sea.

Since the seahorse is so adept at camouflage, it sometimes seeks no cover but lives openly on the seafloor, dun-colored to match.

Problems ahead

An elderly lady recalls swimming as a child with her friends in an inlet of the Red Sea. There were often so many seahorses they would become entangled in the children's bathing costumes, and delighted, the children scooped them up by the handful and took them home to dry and paint with nail varnish and make into earrings. It was another era, when such things were done innocently without thought of endangering species, but this illustrates a time when seahorses were plentiful. Reports now are always of declining numbers.

Seahorses have many enemies. They are wanted alive for the aquarium trade and dead for the esoteric uses of traditional Chinese medicine. But the greatest danger of all is a quietly insidious one, the pollution and destruction of their close-to-shore homes.

The greatest obstacle to protecting seahorses is that we know so little about them. We don't know exactly where they all live, how many there are, how and where different species reproduce, or how they are affected by adverse conditions. If one habitat becomes polluted, do they simply move elsewhere? The only way to find the answer

Dried seahorses were once used as earrings and ornaments, but the practice is now either illegal or frowned upon in many parts of the world.

to these and other questions is to don a cold wetsuit and an oxygen tank and devote untold hours to searching, observing, measuring, and recording. Never let it be said that the Age of Exploration is dead.

Pollution

Because most ocean life lives so close to land, it is bound to be affected by humans. As self-interested as any other animal, we mine, build factories and cities, burn fuel, fertilize fields with noxious pesticides, and pour much of the waste from all these activities into the sea, churning it up in the process. If waste is not directly dumped, it finds its way to the ocean via runoff from the land into rivers, and all rivers empty at last into the sea.

Seahorse homes like reefs, mangrove swamps, estuaries, bays, and seagrass meadows are all poised in the balance between production and destruction, life and death, a balance that is easily tipped. Concentrated untreated sewage is pumped directly into the sea in many parts of the world. This produces a surge of nutrients that cause invasions of new organisms, with the potential to completely destabilize an environment. When coastal land is cleared and developed, or reclaimed, massive amounts of sediment find their way into the sea and cut out sunlight, restricting photosynthesis in single-celled algae, thus strangling the first link in the food chain. Something as simple as hot water, a by-product of power generation, can completely ruin reefs or seaweed forests, which can only survive within a narrow tem-

A Pygmy Seahorse melts into a background of open coral polyps. The delicate polyps retract to protect themselves, but are easily killed by rough treatment. Photo by Tony White.

perature band. Toxins from mining, like lead and mercury, poison whole food chains, with ourselves the ultimate losers.

63

Threat to reefs

Reefs are under dire threat from all directions. Coral grows at a rate of about an inch a year (25mm), so the larger reefs are the work of centuries. Yet it is possible to kill a whole patch of polyps in an instant merely by standing on coral, which tourists frequently do, and by anchoring small boats upon it. But greater damage is caused by local illegal fishing, using homemade bombs which bursts the fishes' air bladders so that they float dead to the surface. Or sodium cyanide is spread, which stuns the reef fish long enough for them to be gathered for the live aquarium trade.

Both of these methods are horribly destructive to reefs. There are laws against them, but reefs often grow in countries which happen also to be very poor, like Indonesia, where half of all reefs are threatened, and the Philippines, where two-thirds are affected. Local fishermen are just trying to scratch a living. It is too tempting a market: 11 million tropical fish are traded annually worldwide, including seahorses. More and more of these are now "grown" commercially, but most still come from the wild.

Commercial firms also mine coral for cement in India, Sri Lanka, and Indonesia, and in addition 375 tons (1.5 million kgs) of coral is harvested each year, again for the

Coral harvesting in the late 1800s: techniques have been modernized, but the practice continues.

aquarium trade. The US has 10 million hobbyists and consumes about one-third of the world's tropical fish and coral exports.

In Australia, it is now necessary to have a license to capture seahorses, and there are

H. kuda, *photographed in the Lembeh Straits of Indonesia. Here cold water wells up from the deep trenches nearby, bringing nutrients that cause some of the inhabitants to grow larger than normal. This kuda is about a third bigger than its normal size. There are rumored plans to dredge the Lembeh Straits for shipping.* Photo by Tony White.

65

Sponges had great commercial value in the nineteenth century but have been mostly replaced by synthetic substitutes. However, they can be badly damaged by bottom trawling fishing tackle

showing people the longer-term economic benefits of preserving reefs for tourism and healthy fisheries.

It must be added that there are also natural dangers to reefs such as outbreaks of disease (though once again this can be triggered by pollutants) and plagues of "crown-of-thorns," *Acanthaster planci*, a huge starfish that inverts its stomach over coral and literally sucks the life out of it. These creatures can each destroy about 60 sq yards (55 sq meters) of reef per year. Sometimes the only way to control them is to put a local bounty on their heads.

Damage by fishing gear

The big commercial fisheries also cause problems. Deep sea trawling nets can be about 260 feet wide (80m), with heavy rubber rollers that crash across the seafloor, smashing up deepwater reefs or shallower seagrass beds as they go. Homes are destroyed, and in addition, for approximately every pound of catch, of flatfish in deeper areas or lobster and shrimp in shallower ones, there are 22 to 44 pounds of "by-catch," unwanted and destroyed marine life.

international conservation groups trying to set up alternative money-making schemes for local communities, with variable success. Education seems to be the best solution,

Many studies are being conducted, and obviously, such fishing methods are under global review.

Potions and poisons

It is not surprising that seahorses were used as medicine in ancient times – just about everything has been pressed into service at one time or another in man's long struggle against disease. Classical texts, and later medieval ones, record remedies such as "ashes of seahorse in wine" to be used as a poison, or mixed with oil of marjoram, as a hopeful cure for baldness. The paunches boiled in wine were thought to produce a crazed adoration of water, an insatiable longing to swim in it, drink it, and hear it flow. Seahorses were also thought to be an antidote for rabies, which was of course widespread then and incurable, and such a grueling disease in humans that one would have tried anything. The second-century Greek natural historian Aelian wrote an anecdote from his wanderings:

"The Hippocampus has been found, by the sagacity of an old fisherman well-skilled in sea matters to be an efficacious remedy. The old man was a Cretan; and his sons, young men, were like their father, fishermen. Well then, it had happened that this old man having caught some Hippocampuses among other fishes, the young men, one after another rendering assistance to the first attacked, were bitten by a mad dog. They were then lying near the shore of Methymna in Crete. The bystanders were pitying them, and directing that the dog should be killed, and his liver given to the young men to eat as a cure for the disease; others were recommending to take them to the temple of Diana and supplicate the goddess for a cure.

The old man, however, praising them for their good advice, took no further notice but proceeded very fearlessly and boldly to cleanse the paunches of the Hippocampus of their contents, and then some of these paunches he roasted and gave them to his sons to eat; others he bruised with vinegar and honey, and applying them as a plaster on the wounds, so got the better of the canine madness in the young men by the longing after water, which you will understand the Hippocampuses excited in them. And in this way he cured the boys, though he was a long time about it."

In need of a remedy –
Seahorses and Traditional Chinese Medicine

"Ashes of seahorse" may seem a thing of the past, but they are not. Today in China, every bustling city and small town across the length and breadth of the country has a traditional medicine shop. The typical shop is small and often very old, lined from floor to ceiling with dark wooden drawers and rows of glass jars crammed with bones, horns, scorpions and lizards, gnarled roots, earth worms, moths – and dried seahorses. There is a mingling of herbal and fishy smells, must and spice. To fill a "prescription," the attendant moves swiftly from drawer to drawer and jar to jar, taking a pinch of herbs here, a cuttlefish bone there, a scoop of gypsum, an insect, slices of ginseng root. He piles the ingredients onto a clean square of paper on the counter and folds it into a neat enticing package, with instructions to chop, steep, boil or make into a soup, a tea, or a poultice. It is all a very far cry from the antiseptic glass and mirrors of a modern pharmacy.

This practice of traditional Chinese medicine has been going on uninterrupted for at least 2000 years. Because of a long-standing distaste for studying anatomy by dissection, knowledge about the human

A dried seahorse ready for the customer at a traditional Chinese medicine shop in San Francisco.

body and what is right or wrong with it had to be obtained by observation, asking questions, taking a range of pulses, and applying the principles of Yin and Yang – a kind of body poetry in which opposites such as heat and cold and light and dark must be balanced to restore health. There are ancient tomes from the Qin, Song, Tang, and Ming dynasties, collections of remedies with

69

Physic Street in old Canton.

names like Peaceful Holy Benevolent Prescriptions.

Seahorses are used to lower cholesterol, aid the mending of broken bones, and treat ailments as disparate as asthma, incontinence, thyroid problems, lethargy, and impotence. There is much made of aphrodisiacs in China and the seahorse is considered to be one. Commonly, a dried seahorse is put in a bottle of alcohol and after a month or so, drunk by men as a virility tonic. Recipes for other problems can be found on the Internet*: for instance, a combination of malytea, scurfpea fruit, walnut seed, tokay, seahorse, dried human placenta, and Chinese magnoliavine fruit has been found to tone up the kidneys.

In Chinese terms, seahorses are quite expensive: on the wholesale market they can fetch as much as $950 per pound, with 150 to 200 individuals in a pound. Only the better off can afford them readily, but China is undergoing a long and steady economic improvement, and demand is increasing.

Traditional medicine is also practiced in Indochina, the Philippines, Malaysia, Indonesia, and Japan, and in far-flung outposts of Chinese culture like San Francisco, using a grand total worldwide of 24.5 million seahorses a year. On the basis of such figures, and reports everywhere of dwindling populations, CITES (Convention on International Trade in Endangered Species of Wild Fauna and Flora, with 161 country members) has placed all species of seahorses on their Red List of Threatened Species. But it is very difficult to prevent the capture of seahorses when there is such a market for them.

Besides seahorses, there are about 80 different kinds of animals or animal parts used in traditional Chinese medicine, including leopard bones, tortoise shells, rhino horns, shaved ivory, bear's gall, and tiger's penises. Many of these are thought to be aphrodisiacs. Perhaps patients in time will be lured away by legends of Viagra. The seahorses live in hope.

*www.sungardenherbs.com

Fact, fiction and fantasy

The boundary between myth and fact was once much hazier than it is today. In the medieval telling of history, kings and battles came alive on the same page as dragons and miracles, and in natural history, two quite different seahorses were taken equally seriously – the delicate bony creature washed up on the beach after a storm, and the "sea-monster," Hippocampus, from classical poetry and the pantheon of the gods.

In 1558, a Swiss physician and scholar called Konrad Gesner gathered every then-known fact (and fiction) about animals, and published in Latin his fabulously illustrated *Historiae Animalium*. He has two entries for seahorses, one for the mythical, and one for the real creature. The illustration for an actual seahorse seems to be drawn from a dried specimen and is classed as a "marine insect"; the sea-monster is a vigorous beast called Equo Neptuni, with fins instead of hooves on its rearing front legs, and a thick, muscular serpent's tale.

A hundred years later in England, Edward Topsell compiled another all-encompassing natural history, borrowing heavily from Gesner. He named it, with a flourish: *A Historie of foure-footed Beasts and*

Seahorses have such appeal they can even sell wine.

Seahorse-inspired kitsch: trinkets and children's stickers.

The arms of the port city of Newcastle-upon-Tyne is flanked by mythical seahorses.

 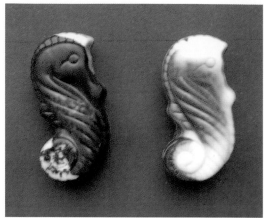

Guylian, makers of sea-inspired chocolates, have recently sponsored a major study of seahorses.

Serpents Describing the true and lively figure of every beast . . . their love and hate to mankind. But fact and fiction still being in a glorious muddle, his seahorse entry is a colorful description of the "filthy" hippopotamus!

Is it far-fetched to speculate that the seahorse, with its unique part-horse, part-fish, part-reptilian appearance, might have sewn a seed for demi-creatures like mermaids and centaurs? The seahorse was apparently thought of as the baby form of a mythical steed that lived in the ocean. Ancient Greeks, their minds filled with poetry, dreamed up the Naiads, 50 beautiful sisters in Poseidon's retinue who rode naked, bareback, on seahorses and dolphins, rescuing drowning sailors in the Mediterranean. The Romans transformed Poseidon into the sea-god Neptune, and to this day seahorses pull his chariot all over Europe, with wild eyes and swirling manes sculpted in stone.

Seahorses, both the real and the mythical, have been used as insignia for hundreds of years, especially on coats of arms for naval families or port cities where there is a vested interest in the sea, and as military emblems. They were embroidered on the cloaks of Qing Dynasty generals in China, and are used today in the American military. There is something of bounty and good omen in the image of this creature. Captain Cook's ship, the *Resolution*, in which he sailed the Pacific on his second voyage, ploughed the waves with a carved seahorse at its prow.

Countless companies use the seahorse as an emblem, from winemakers to finance houses. Alas, they are such sweetly benign creatures that they have been almost entirely usurped by purveyors of bad taste: witness the countless lurid seahorse ornaments for sale in beach resorts.

The seahorse in various fanciful forms is continually used as a military emblem. This is a badge from a US air squadron, now disbanded, that used VA-55 Warhorses to bomb Libya in 1986.

In reality, seahorses always were and still are rare little jewels. In *Curiosities of Natural History* (1883), Frank Buckland brings alive the newness and excitement of seahorses in a still-to-be-explored-world where few traveled far from home: "A pair of these curious and pretty little creatures were lately exhibited alive in the Regent's Park Gardens. They were brought from the mouth of the Tagus, and presented by a Portuguese gentleman. Two were offered not long ago, alive, in Hungerford Market, for a guinea each."

A heraldic seahorse.

75

Reports of narwhales inspired the artist, J. Hoefnagel, to create a horned seahorse in a painting c. 1600 from the Bestiary of Emperor Rudolph II of Austria.

A lovingly observed painting of a seahorse, brought in live by a fisherman in 1836. From R. T. Lowe's History of Fishes of Madeira *(1860).*

The loveliest of tributes comes from R.T. Lowe's *History of Fishes of Madeira* (1860), which describes how a single female was brought alive and "in full vigour" by a fisherman, as a great curiosity, on the 17th of August, 1836. It had been taken entangled in his net or lines, about five hundred yards off the Loo Rock, or Ilheo, and lived in a glass of seawater after its capture for more than twelve hours. Lowe transcribes the on-the-spot description, giving it as much respect as he does his dry scientific observations.

Clockwise from left, illustrations from Konrad Gesner's sixteenth-century tome: a real hippocampus, Neptune's *seahorse, and an altogether different sort of "seahorse" - a hippopotamus in the process of devouring a crocodile.*

A series of British stamps depicting Britannia being drawn by a team of seahorses was issued between 1913 and 1937; and in Bermuda, His Majesty himself drives the chariot.

"Through the lens, the sides appear most beautifully speckled with minute orange dots on a livid-blue ground; and the whole surface of the head and body are spangled with pure opaque-white scattered dots or points . . . The eyes are like some brilliant jewel set in mosaic. The iris, as in the chameleon, is most beautifully painted with alternate rays or tessellated bars of glittering garnet-red and dazzling white on the inner edge next the pupil; the circumference being thickly studded with the brightest pure white dots. The pupil is black, but reflects in rapidly succeeding variation, as it moves, most brilliant topaz, straw-color, or brassy tints, in different lights."

Lowe paints for us in words a picture of the moving, breathing creature that no photograph or painting or illustration can quite capture. He reminds us that the seahorse, like all living things, is not just an object to capture in an aquarium or use for medicine, nor an environmental statistic to be set to rights, but a unique creation whose perfect and elusive beauty makes our lives the more wonderful.

Further Reading

Seahorses, Pipefishes and their relatives: A Comprehensive Guide to Syngnathiformes
by Rudie H. Kuiter, 2002

Seahorses: an identification guide to the world's species and their conservation
by S.A. Lourie, A.C.J. Vincent and H.J. Hall, 1999

The Sea-Horse and its Relatives
by Gilbert Whitley and Joyce Allan, 1958

A Treasury of New Zealand Fishes
by David H. Graham, 1953

Review of Hippocampus by I. Ginsburg
Proceedings of the U.S. National Museum, 1937

"The Curious Life Habits of the Sea Horse" by Rene Thevenin
Natural History, March 1936

Flora e Fauna del Golfo di Napoli (Monograph 36)
by M. Rauther, 1925

The Royal Natural History, 6 volumes
R. Lydekker, 1896

History of Fishes of Madeira
by Richard Thomas Lowe, 1860

HLOOW 597
.6798
W214

WALLIS, CATHERINE
SEAHORSES

LOOSCAN
03/07